i-SPY

in the night sky

SPY IT! SCORE IT!

Introduction

To appreciate the sky and stars in their full glory you will need a dark night and clear sky. You can see most when you are in the country or at the seaside, or maybe at the bottom of your garden.

Here are some top tips: Always tell your parents if you want to watch the stars. Ask your parents or an elder sibling to come with you. Always wear warm clothes – even in the summer, it can be cold at night. And take a torch so you can see your book. Bright lights dazzle your eyes though, so cover the end of the torch with red tissue paper or red plastic film to give a dim red light that doesn't spoil your night vision!

How to use your i-SPY book

Keep your eyes peeled for the i-SPYs in the book.

If you spy it, score it by ticking the circle or star.

50 POINTS

Items with a star are difficult to spot so you'll have to search high and low to find them.

If there is a question and you know the answer, double your points. Answers can be found at the back of the book (no cheating, please!)

Once you score 1000 points, send away for your super i-SPY certificate. Follow the instructions on page 64 to find out how.

The Moon illusion

When the Moon (especially the Full Moon) is low on the horizon it appears enormous. In fact, it is actually exactly the same size as when it is high in the sky. This is known as the 'Moon Illusion' – your brain simply thinks it is bigger when compared with distant hills or buildings. Test it for yourself: hold a finger at arm's length and you will see that it is about twice the width of the Moon, wherever the Moon is in the sky. Actually, the same effect occurs with the Sun, but remember you must never look at the Sun through binoculars or any telescope because of the danger of damaging your eyes.

10 POINTS *See the Moon low on the horizon.*

See the Moon high on the horizon. **10 POINTS**

Learning about stars

Visit a planetarium

A good way of learning about the sky is to visit a planetarium, wher a special projector shows the night sk on the inside of a dome.

10 POINTS

Learn stars in a planetarium

Planetariums have different programmes about the planets, galaxies and black holes, for example. You can learn about how to find your way around the sky. With the naked eye you can see about 8500 stars over the whole sky.

10 POINTS

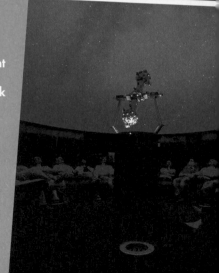

Use a planisphere

Another good way of finding out which star can be
seen, and at what times, is to use a planisphere. This
is a special map of the sky, marked so that you
can set it to any date and time. It then shows
exactly what you will be able to see. Digital
versions are available online.

15 POINTS

Telescopes

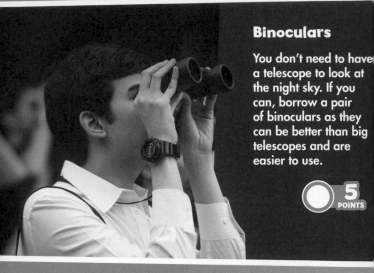

Binoculars

You don't need to have a telescope to look at the night sky. If you can, borrow a pair of binoculars as they can be better than big telescopes and are easier to use.

5 POINTS

Refracting telescope

A refracting telescope uses lenses at each end of a tube. The large one where the light enters is known as the objective, and the smaller one is called the eyepiece.

10 POINTS

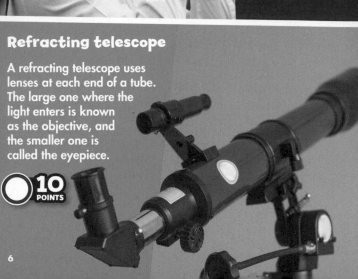

Reflecting telescope

Another type of telescope, known as a reflector, uses mirrors instead of lenses (as well as an eyepiece), and you have to look in the side of the tube. The mirrors act as a magnifier to enlarge the subject.

Schmidt-Cassegrain telescope

Some telescopes use both lenses and mirrors. They are popular because they are easy to carry. One type you may see is called a Schmidt-Cassegrain, named after two astronomers who designed telescopes.

The rotating sky

How the sky rotates

Although the stars are actually scattered about in space they look as if they are on the inside of a giant dome. When you stand outside – like the little figure in the diagram below – the stars are overhead and the horizon is all around you. Just as the Sun moves across the sky during the day, the stars move during the night, because the Earth itself is rotating. One star, called Polaris, appears to stand still at the North Pole of the sky, and all the other stars seem to rotate around it. Only in a photograph, such as the one shown to the right, can you see that Polaris is not quite at the North Pole. The exposure needed for the photograph was several hours, and all the stars have left trails as the sky moved. The very short, bright trail near the centre was made by Polaris.

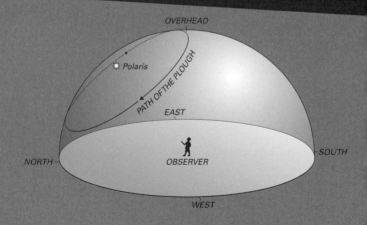

OVERHEAD

Polaris

PATH OF THE PLOUGH

EAST

NORTH

OBSERVER

SOUTH

WEST

15 POINTS

What is another name for Polaris?

The rotating sky

Find the Plough and Polaris

How do you find Polaris? First you find a well-known group of seven stars known as the Plough (or Dipper in North America), with an easily remembered shape, rather like a saucepan with a long handle. Because they are very close to the North Pole, the stars of the Plough never dip below the horizon. The two end stars of the Plough are called the Pointers, because a line through them points to Polaris.

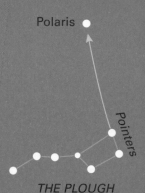

ssiopeia

Pictures in the sky

Because the Earth revolves around the Sun once a year, the Sun appears to move slowly across the sky, hiding different stars during the daytime as the months go by. So the stars that you see at night also change with the time of the year. Later in this book you can see what the sky looks like during each of the four seasons. To help them find their way around the sky at different times of the year, earlier pioneers invented names for various groups of stars, or constellations. Astronomers use the Latin names for constellations in modern star atlases, so everyone, anywhere on Earth, knows which groups of stars they are talking about.

10 POINTS *Old book of constellations*

10 POINTS *How many constellations cover the whole sky?*

10 POINTS *Modern star atlas*

The names of stars

Most bright stars have individual names, and many of these come from Arabic words that described their positions in the imaginary constellation patterns in the sky. The red star at the top left of Orion, for example, is known as Betelgeuse, which probably means 'the woman in the middle' (the Arabs saw this constellation as a woman). The brilliant bluish star at the bottom right corner is Rigel, meaning 'foot'. The names Polaris and Ursa Major are, however, from Latin.

Orion

10 POINTS *What is the English name for Orion?*

10 POINTS *Book of constellations*

10 POINTS *Orion*

Rising and setting stars

Constellations close to Polaris (mainly those shown on the chart on page 25) always remain above the horizon. (They are known as 'circumpolar constellations'.) Many other constellations are farther away from Polaris and the North Pole, so they rise and set during the night. The diagram shows how the constellation called Orion (page 36) rises above the horizon in the east, becomes highest in the south, and then disappears as it sets in the west.

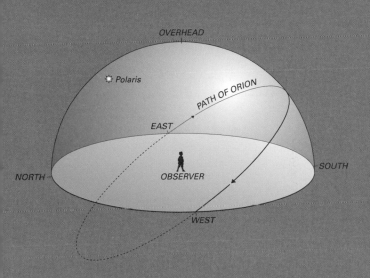

The planets

Watching planets

Stars shine because they are extremely hot and give out their own light. The Moon and planets, on the other hand, only reflect light from the Sun. Two planets, Venus and Jupiter, shine like very bright stars, and two others, Mars and Saturn, although fainter, are also sometimes easy to see. All these planets (and the other, fainter ones) orbit the Sun just like the Earth.

Saturn (a yellowish colour)

Mars

Jupiter

Venus

Venus and Jupiter with the Moo

Jupiter's satellites

Of Jupiter's many satellites in orbit around it, four of them are large (three are larger than our Moon). They are easily seen through binoculars, appearing on each side of the planet.

Who first saw Jupiter's satellites through a telescope?

Jupiter's satellites

Movement of the planets

Because they are orbiting the Sun, planets do not have fixed positions like the stars, so they cannot be drawn on the maps here. You can sometimes find details of where they are in the newspapers and also on the internet. Venus is always close to the Sun, so it is in the west in the evening, and in the east early in the morning.

 There are several known dwarf planets, the largest of which is Pluto (diameter 2374 kilometres). All except Ceres orbit in the outermost Solar System.

The Moon

The phases of the Moon

As the Moon goes around the Earth, we see it by the light from the Sun that it reflects. It always turns the same side towards us, so sometimes it is in darkness, sometimes partly lit, and sometimes a complete, bright circle. At New Moon the disk is completely dark, but a very thin, young crescent that appears in the western sky shortly after sunset is often called by the same name. The ends of the crescent (the horns) are turned towards the east. As the days go by, the Moon appears farther to the east at sunset, and the bright area waxes (increases). When half of the Moon can be seen, it is called First Quarter. The bright area continues to increase, until the whole face of the Moon is shining at Full Moon. Between First Quarter and Full Moon, the Moon is called waxing gibbous. After Full Moon the bright area begins to wane (decrease), first through waning gibbous, then Last Quarter, and finally to waning crescent, until it disappears close to the Sun in the east at sunrise.

 10 POINTS Narrow crescent moon (New Moo

 10 POINTS The horns the cresce

 10 POINTS First Quarter

 10 POINTS Waxing gibbous

 10 POINTS Full Moon

 10 POINTS Waning gibbous

 10 POINTS Last Quarter

 10 POINTS Waning crescent

Waxing phases

Full Moon

Waning phases

The Moon

Earthshine

Sometimes when the Moon appears as a crescent, you can see the rest of it, faintly illuminated by sunshine that has been reflected from the Earth on to the dark part of the Moon. This is called Earthshine.

Lunar seas and highlands

Some parts of the Moon's surface are darker than others. These dark grey areas are called maria (Latin for 'seas'). They are flat lava plains. Most of the Apollo missions to the Moon landed in the maria, because they were safer and more predictable than the more rugged areas. The lighter areas are called the highlands, and they contain nearly all the craters that pockmark the face of the Moon. One Apollo mission, Apollo 17, landed on the edge of the highlands. At Full Moon, some craters show bright rays where powdered rock has been splashed out by the impacts. The longest rays belong to a crater called Tycho.

 Maria

 Highland

 Moon landings
Only six NASA space missions have landed on the surface of the Moon:

Apollo 11 – 16 July 1969   Apollo 15 – 26 July 1971
Apollo 12 – 14 November 1969   Apollo 16 – 16 April 1972
Apollo 14 – 31 January 1971   Apollo 17 – 7 December 1972

Satellites, comets and meteors

Artificial satellites

As you watch the sky, especially at dawn and dusk, you will often see a slow-moving point of light crossing the sky. This is an artificial satellite, orbiting the Earth. Some, such as Iridium satellites, become very bright and then fade away.

10 POINTS

Comets

You may be lucky enough to see a comet. Only rarely do these become bright enough for you to see them with the naked eye. A comet is a mixture of ice and dust – meteor dust comes from comets – and the heat of the Sun often makes it grow a long tail.

50 POINTS

TOP SPOT!

Meteors

Sometimes you see a sudden, short streak of light. This is often called a 'shooting star', but its proper name is a meteor. (They are not stars, but tiny grains of dust that are burnt up as they dash through the Earth's atmosphere.) On a few nights of the year, the Earth passes through a cloud of dust, and we have a meteor shower, with dozens of meteors each hour. Very rarely, a meteor is brighter than any of the stars or planets; it is then called a fireball.

50 POINTS Fireball

TOP SPOT!

30 POINTS Meteor shower

10 POINTS Meteor

i Halley's Comet is the most famous comet. It returns every 76 years, most recently in 1986.

The Great Bear

It is best to start learning the sky with the Plough (page 8). It forms just part of the constellation of Ursa Major, the Great Bear. The picture shows how the constellation used to be drawn on old maps of the sky. The two end stars are called the Pointers, because a line through them points to Polaris (page 8), the star that stands still. The Pointers are called Dubhe and Merak.

 5 POINTS   Ursa Major

 10 POINTS   Dubhe

 10 POINTS   Merak

 10 POINTS   Why do the Grea and Little Bear appear odd in al old constellation drawings?

The Little Bear

Polaris is the brightest star in Ursa Minor, the Little Bear. Another bright star in this constellation is Kochab, which is one of a pair of stars called The Guards. The diagram shows how the two constellations appear to swing round the North Pole (and Polaris) during the course of the night.

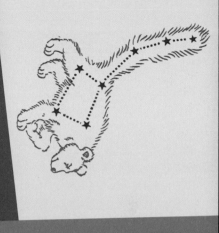

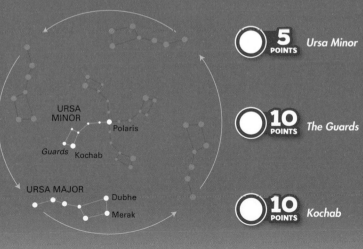

URSA MINOR

Polaris

Guards Kochab

URSA MAJOR

Dubhe

Merak

5 POINTS *Ursa Minor*

10 POINTS *The Guards*

10 POINTS *Kochab*

NORTH

Constellations near the North Pole

Cassiopeia

There are several other constellations near the North Pole. If you find the star called Megrez, where Ursa Major's 'tail' joins the body, and imagine a line from there to Polaris and beyond, you come to a group of five stars in the shape of a 'W' – or an 'M' when they are the other way up! These form the constellation of Cassiopeia (page 25), an ancient, legendary queen.

Draco

Snaking its way around the pole is a long, faint constellation called Draco, the Dragon. At the end of its long, winding 'body', four stars make up its lozenge-shaped 'head'. The middle star of the three that form Ursa Major's 'tail' is Mizar. Between Mizar and The Guards is Thuban in Draco.

Cepheus

Between the head of Draco and Cassiopeia is the constellation of Cepheus (page 39), the legendary king and husband of Cassiopeia. The stars in this constellation form a pointed shape like the gable end of a house.

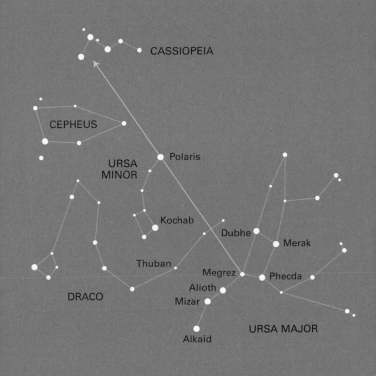

CASSIOPEIA

CEPHEUS

URSA
MINOR

Polaris

Kochab

Dubhe

Merak

Thuban

Megrez

Alioth

Phecda

Mizar

DRACO

Alkaid

URSA MAJOR

 Megrez

 Thuban

 Mizar

 What is strange about Mizar (look at it through binoculars if you are not sure)?

A legend in the sky

Andromeda, Pegasus and Perseus

Do you know the Greek legend of how Andromeda, the daughter of Cassiopeia and Cepheus, came to be chained to a rock, and was about to be killed by the terrible sea monster Cetus (see map page 48)? She was rescued just in time by Perseus, who had earlier killed the dreadful Gorgon, Medusa. Medusa's head, with snakes for hair, turned anyone who looked at it to stone. Perseus was clever and only viewed her reflection. He showed the head to Cetus, who became a rock. After rescuing Andromeda, Perseus and Andromeda rode away on Pegasus, the flying horse (which appears upside down in the sky). You can find all these legendary beings in the sky (see the maps on pages 38 and 48 for examples). The old constellation drawing of Perseus shows him holding Medusa's head. This is represented in the sky by the star Algol, whose name comes from the Arabic words Al Guhl, meaning 'the demon'. The photograph shows the constellation pattern and a bright comet, Comet Hyakutake.

10 POINTS *Algol*

Comet Hyakutake

 10 POINTS *Cetus*

10 POINTS *Pegasus*

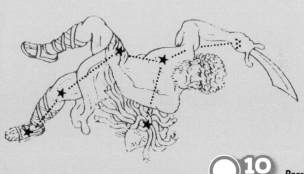

10 POINTS *Perseus*

The Summer Triangle

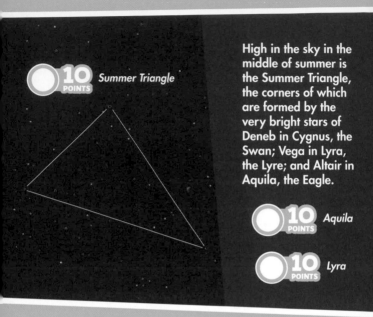

High in the sky in the middle of summer is the Summer Triangle, the corners of which are formed by the very bright stars of Deneb in Cygnus, the Swan; Vega in Lyra, the Lyre; and Altair in Aquila, the Eagle.

10 POINTS *Summer Triangle*

10 POINTS Aquila

10 POINTS Lyra

Cygnus

Cygnus is sometimes known as the 'Northern Cross' and the name Deneb means 'the tail' in Arabic because it marks the tail of the Swan. The bright star at the other end (the 'beak') of the constellation is Albireo, but its name has no real meaning, because it arose after a series of errors and bad translations.

10 POINTS Cygnus

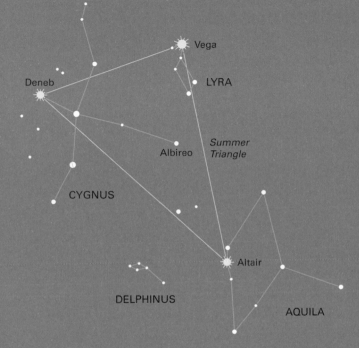

Vega

LYRA

Deneb

Albireo

*Summer
Triangle*

CYGNUS

DELPHINUS

Altair

AQUILA

 10 POINTS *Altair*

 10 POINTS *Vega*

 10 POINTS *Deneb*

 10 POINTS *Albireo*

29

The Milky Way

10 POINTS

The Milky Way is a silvery band of light that stretches right across the sky and is easiest to see in summer.
The faint light comes from hundreds of thousands of stars which are so closely crowded together that you cannot see them as individual stars. With binoculars you can see that there appears no limit to the number of stars in the Milky Way, which is actually our edge-on view of the main disk of the Galaxy in which we live, and which contains over one hundred thousand million stars.

Deneb

Vega

LYRA

The Great Dark Rift

CYGNUS

Altair

AQUILA

SAGITTARIU

Great Dark Rift

The densest part of the northern Milky Way runs from Cygnus down towards Sagittarius (page 45). Cygnus appears to be flying down the Milky Way. Can you see the Great Dark Rift in Cygnus? This may look like a 'hole in the stars', but actually there are just as many there as elsewhere in the Milky Way, but most are hidden by dark clouds of dust between the stars.

20 POINTS

> *i* Black holes are completely invisible, because not even light can escape from them, but they can be detected by their effects on nearby stars. The closest known black hole is orbiting a faint star in Cygnus.

The Milky Way

Stars and clusters

Open clusters

Inside our own Galaxy stars are born in groups (known as clusters) from the material between the stars. Those with young, blue stars are known as open clusters, and usually contain tens of stars which are only a few million years old. The most famous open cluster is the Pleiades, in Taurus (see page 38). Other open clusters are older, such as the Double Cluster in Perseus (see map on page 43), which can be seen through binoculars.

10 POINTS

Pleiades

Pleiades

Globular clusters

There are other clusters that are much older, at least ten thousand million years old. These are the globular clusters, which are spherical in shape and each contain many thousands of stars. The most famous in the northern sky is M13, in Hercules (see map page 51).

M13, in Hercules

M13, in Hercules

15 POINTS

Orion Nebula

Nebulae

A cloud of gas between the stars is known as a nebula (plural 'nebulae'), but only one, the great Orion Nebula (page 37), is easily visible to the naked eye.

20 POINTS *Orion Nebula*

The Northern sky

The seasonal charts

The charts on the following pages show the sky during the four different seasons of the year. There are two charts for each season, one looking east, the other looking west.

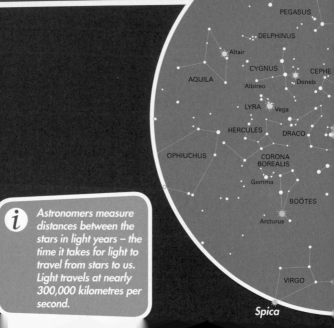

AQUARIUS

PEGASUS

DELPHINUS

Altair

CYGNUS

CEPHE

AQUILA

Albireo

Deneb

LYRA · Vega

HERCULES

DRACO

OPHIUCHUS

CORONA
BOREALIS

Gemma

BOÖTES

Arcturus

VIRGO

Spica

i Astronomers measure distances between the stars in light years – the time it takes for light to travel from stars to us. Light travels at nearly 300,000 kilometres per second.

Stars around the North Pole

This chart shows the key stars you can see if you live in the Northern Hemisphere. With it you can find your way from one constellation to another. The North Pole of the sky and Polaris are in the centre.

PISCES

CETUS

Mira

ROMEDA

Hamal

ARIES

TAURUS

TRIANGULUM

Algol   *Pleiades*

PERSEUS   Aldebaran

Rigel

OPEIA

Capella

is

Betelgeuse   ORION

R

AURIGA

Castor

GEMINI

Pollux

A MAJOR

Procyon

CANIS MINOR

CANCER

LEO

Regulus

ebola

HYDRA

Alphard

Light from Sirius (see next page), the brightest star in the sky, takes 8.6 years to reach the Earth.

Winter - looking East

Winter is the best time to see Orion and the stars of Canis Major and Canis Minor, the Greater and Lesser Dogs. Betelgeuse, in Orion, is a gigantic type of star, called a red supergiant. If it were in the Solar System in place of the Sun, the Earth would be below its surface. Its colour is in sharp contrast to bluish-white Rigel on the other side of the constellation. Sirius, in Canis Major, is the brightest star in the sky. Venus and Jupiter, which are planets (page 14), are the only objects that ever appear brighter.

15 POINTS The stars of Orion's belt

10 POINTS *What is another name for Sirius?*

10 POINTS *Betelgeuse*

The Orion Nebula is a vast cloud of gas and dust, just visible to the naked eye as a fuzzy spot below Orion's belt. Dozens of new stars are being created within it.

10 POINTS *Rigel*

10 POINTS *Canis Major*

10 POINTS *Sirius*

10 POINTS *Canis Minor*

Winter – looking West

Altair, in the Summer Triangle (page 28) will have disappeared, but Vega can still be seen low on the horizon in the north-west, while Deneb is higher up in the same part of the sky. Andromeda, with its great galaxy (page 49) and the small neighbouring constellations of Triangulum, the Triangle, and Aries, the Ram, with one bright star, Hamal, are still easy to see. High in the south is the constellation of Perseus, and lower down, Taurus, the Bull, whose 'eye' is orange Aldebaran. Taurus contains two clusters, the Pleiades (page 32) and the 'V' of stars near Aldebaran, known as the Hyades.

Hyades
15 POINTS

Double cluster
20 POINTS

Aldebaran
10 POINTS

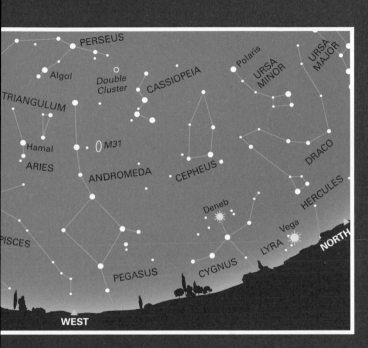

10 POINTS *Hamal*

10 POINTS *Taurus*

10 POINTS *Aries*

10 POINTS *Triangulum*

Spring – looking East

If you follow the line of the Pointers, not towards Polaris, but in the opposite direction, you come to the great constellation of Leo, the Lion. The 'backward question mark' of stars (with Regulus as the 'dot') forming the lion's head (see the next chart on page 42) is known as The Sickle. At the other end of Leo, Denebola forms the 'tail' of the lion. You can follow the curve of the tail of Ursa Major, or a line from Regulus to Denebola, to help you find Arcturus, the brightest star in Boötes, the Herdsman. Low on the horizon are the four stars of Corvus, the Crow.

 Arcturus

 Corvus

 Boötes

 Denebola

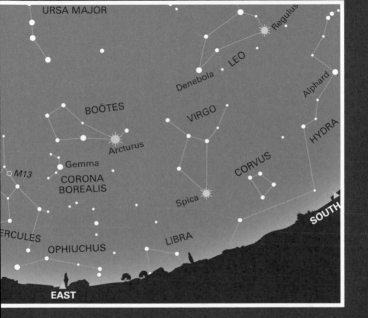

 Leo

 Regulus

 Pointers

 The Sickle

Spring – looking West

Spring is a good time to see the constellation of Gemini, the Twins, which consists of two lines of stars running west from the two bright stars Castor and Pollux. Between Leo and Gemini is the faint constellation of Cancer, the Crab. Below Gemini is the small constellation of Canis Minor, the Lesser Dog, with just one bright star, Procyon. High in the north-west is the constellation of Auriga, the Charioteer. In old drawings he is shown carrying two young goats in his arms, so the little triangle of stars to the right of bright Capella, is known as The Kids.

 The Kids

 Cancer

 Auriga

 Capella

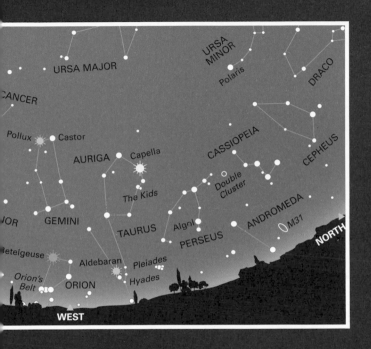

 Castor

 Pollux

 Gemini

 Procyon

Summer – looking East

Summer is the best time to see the great Summer Triangle (page 28) and the star clouds of the Milky Way, which consist of thousands of stars, too close together to be seen as separate points of light. Very low in the south is Sagittarius, the Archer. Can you spy the 'Teapot' among its stars? To the east is the fainter constellation of Capricornus, the Sea-Goat, and the rather brighter one of Aquarius, the Water Carrier. Between them and Cygnus lies the tiny constellation of Delphinus, the Dolphin.

URSA MINOR
Polaris
URSA MAJOR
CEPHEUS
PERSEUS
CASS.
AURIGA
Double Cluster
Capella
M31
NORTH
Algol
ANDROMEDA

Aquila

20 POINTS 'Teapot'

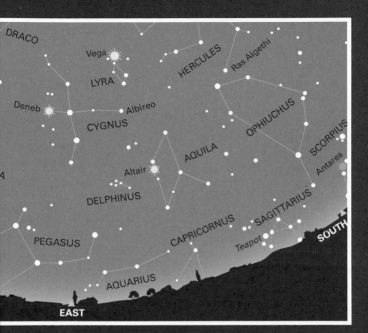

Aquarius

Delphinus

Capricornus

Sagittarius

Summer - looking West

If you follow the curve of the tail of the Plough (as described on page 40) and go beyond Arcturus, you come to Spica, the brightest star in the constellation of Virgo, the Virgin. East of Boötes is the semicircle of stars of the constellation of Corona Borealis, the Northern Crown. The brightest star is called Gemma, the Jewel. To the east lies the large constellation of Ophiuchus, the Serpent Bearer. Below this is Scorpius, the Scorpion. Its brightest star is Antares, the 'Rival of Mars', so called because, like the planet, it is deep red in colour. Between Antares and Spica lies the constellation of Libra, the Balance (or Scales).

 Antares

 Gemma

 Corona Borealis

 Libra

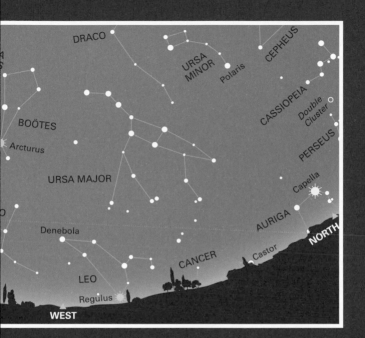

 Ophiuchus

 Spica

 Scorpius

 Virgo

The Northern sky

Autumn – looking East

Rising high in the east is the Great Square of Pegasus. The star at the north-eastern corner actually belongs to Andromeda. Part of the way along the line of four bright stars in Andromeda, two more point at right angles to the north, indicating the position of the Great Andromeda Galaxy (also known as M31). It needs to be very clear and dark for you to see this galaxy. Just think, the light you see started its journey at about the time when the first humans began to evolve from ape-like creatures! Below Pegasus lies the faint constellation of Pisces, the Fishes, and still farther south, there is the large area of sky known as Cetus, the Sea Monster. Unfortunately many of the stars in this area are rather faint. If you are lucky, however you may just be able to see, very, very low in the south, Fomalhaut in Piscis Austrinus, the Southern Fish.

20 POINTS The Great Andromeda Galaxy

15 POINTS Great Square of Pegasus

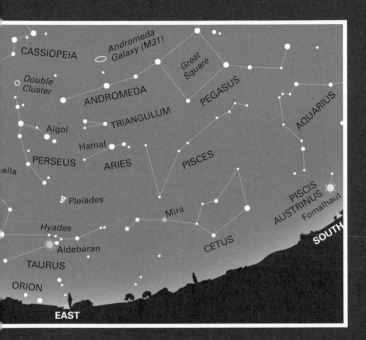

CASSIOPEIA
Andromeda Galaxy (M31)
Double Cluster
Great Square
ANDROMEDA
PEGASUS
TRIANGULUM
AQUARIUS
Algol
Hamal
PERSEUS
ARIES
PISCES
ella
Pleiades
PISCIS AUSTRINUS
Fomalhaut
Mira
Hyades
Aldebaran
CETUS
SOUTH
TAURUS
ORION
EAST

10 POINTS — *Andromeda*

10 POINTS — *Fomalhaut*

10 POINTS — *Cetus*

10 POINTS — *Pisces*

Autumn – looking West

Like the constellations of Andromeda and Pegasus, Hercules, the legendary hero, is upside down! One foot rests on Draco in the north. The name of one star Ras Algethi, means 'the kneeling man's head' in Arabic. The main part of the body is known as the 'Keystone' because the arrangement of the four stars is just like the shape of the central stone in an archway. On the western side of the Keystone lies the globular cluster, M13 (see page 33). It is still possible to see most of the Summer Triangle and the bright stars of Vega, Deneb and Altair (see page 28), as well as much of the Milky Way which also runs through the constellations of Cassiopeia and Perseus.

How many labours did Hercules have to accomplish?

The cluster M13

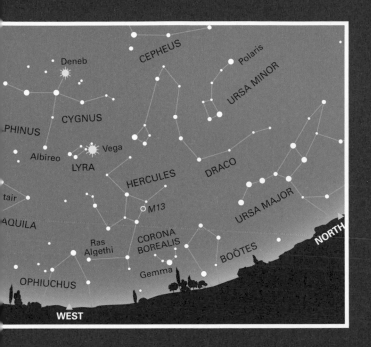

10 POINTS *Ras Algethi*

10 POINTS *Hercules*

Eclipses of the Moon

How do eclipses of the Moon happen?

Sometimes, at Full Moon (page 16), the Earth comes between the Sun and the Moon and casts its shadow on the Moon, giving a lunar eclipse. (It does not always do this, because the Moon is sometimes above and sometimes below the line connecting the Earth and the Sun.) The Moon moves through the shadow from west to east, and you can watch the shadow creep across the Moon. The changes are easy to see with binoculars.

Total lunar eclipse

When the Moon does pass through the centre of the shadow, the whole surface is darkened in a total eclipse, which may last as long as 100 minutes. Very occasionally, if there are lots of clouds or large amounts of volcanic dust in the Earth's atmosphere, the Moon may appear very dark and almost disappear.

It is thought that the Moon was created when a body, the size of Mars, hit the Earth with a glancing blow. Fragments formed a disk around the Earth and later combined into the Moon.

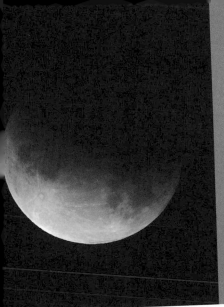

Partial lunar eclipse

Because of the variations in the Moon's position, frequently it does not pass through the centre of the Earth's shadow. Part of the Moon – usually the northern or southern region – remains sunlit, giving a partial eclipse.

20 POINTS

The colour of the Moon

Generally the Moon does not disappear completely at total eclipse. The Earth's atmosphere lets some red light through, so the Moon appears red. Usually there is enough light to make out the general shape of the dark maria (page 18), and sometimes even some prominent craters.

10 POINTS

Solar eclipses

How do solar eclipses happen?

Although it does not happen at night, the Moon sometimes comes between the Sun and the Earth (at New Moon), similar to the way lunar eclipses occur. The Sun and Moon appear almost exactly the same size in the sky, and the Moon blocks out the sunlight to give a solar eclipse. This is a sight not to be missed if you have the chance.

Total solar eclipse

Along a very narrow line on the surface of the Earth, the Sun may be completely hidden by the Moon, in a total solar eclipse, generally for no more than about 3–4 minutes. The way the shadow sweeps across the Earth is very dramatic, and during totality the beautiful outer atmosphere of the Sun, the corona, becomes visible.

50 POINTS

TOP SPOT!

← Progression of eclipse

You must NEVER look at the Sun through binoculars or a telescope, because it would damage your eyes or even blind you.

It takes light about eight minutes to reach us from the Sun, but about four years to travel to the nearest star (which is called Proxima Centauri).

Partial solar eclipse

The Moon only throws a tiny shadow on the surface of the Earth, so you are much less likely to see a solar eclipse than a lunar one. You are most likely to see a partial solar eclipse, when the Moon hides just part of the Sun.

20 POINTS

Annular solar eclipse

Because the distances to the Moon and Sun change, sometimes the Moon does not completely cover the disk of the Sun, and a ring of light remains visible, in an annular eclipse. Although only visible from a narrow path on the Earth, they can last for about 12 minutes, and are slightly more common than total eclipses.

TOP SPOT!

40 POINTS

Clouds at night and aurora

TOP SPOT!

Nacreous clouds

Rarely, shortly after sunset (or before sunrise, if you are up that early), you may see clouds with beautiful bands of different colours. These are nacreous (mother-of-pearl) clouds made of ice particles. They are so high (15–30km) that they are illuminated by the Sun, even though the ground is in darkness.

30 POINTS

Noctilucent clouds

Occasionally, during the summer months of late May, June and July, bluish-white, wispy clouds may be seen in the north, around midnight. These are noctilucent ('night-shining') clouds, the highest of all at about 80km above the ground. They are also made of ice, and reflect light from the Sun that is hidden far below the northern horizon.

30 POINTS

TOP SPOT!

Aurora

Quite frequently, the Sun sends out streams of charged (electrified) particles. When these hit the Earth's atmosphere, they cause the air to glow, giving an aurora: changing patterns of light high in the sky.

Observatory

Observatory

You may be lucky enough to visit an observatory, which is where professional astronomers study the stars. They often have several telescopes, which are usually very large reflectors (page 7). Do not expect to look through one, though! The telescopes are only used with highly complicated digital cameras and other equipment. But many observatories have visitor centres for the public.

 10 POINTS *Observatory*

 10 POINTS *Observatory visitor centre*

 10 POINTS *Observatory telescope*

Royal Greenwich Observatory

Amateur observatory

Many local amateur astronomical societies have observatories, and these often hold open days where you can visit and may be able to look through some of the telescopes. Like professional observatories, the telescopes are usually protected by a dome.

 10 POINTS Amateur observatory

 10 POINTS Dome

 10 POINTS Look through a telescope

Radio observatory

Nowadays, astronomy is carried out in many different ways: with radio telescopes (with both large dishes and 'TV-type' aerials), and with satellites. Again, some radio observatories have visitor centres, and you can see typical satellites at some space centres.

 20 POINTS *Radio observatory visitor centre*

 10 POINTS *Radio observatory*

 10 POINTS *Dish telescope*

Jodrell Bank Observatory

A voyage of discovery

You can learn a lot about astronomy by visiting a museum. Some are in old observatories. There you will often find displays explaining about planets, stars, galaxies, and other strange things like black holes! But you can also see many of the old instruments that earlier astonomers used.

Celestial globe

One of the simplest devices is a celestial globe which shows the stars, usually with the old constellation drawings.

10 POINTS

Astrolabe

A more complicated device is an astrolabe, which is a special form of sky map. A planisphere (page 5) is a simplified modern version.

10 POINTS

Astronomy in museums

Armillary sphere

You may see an armillary sphere, which is a collection of rings that may be used to show the movements of the Sun, Moon and stars in the sky.

 10 POINTS

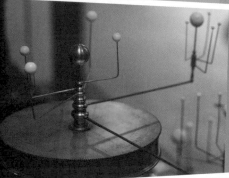

Orrery

An orrery is a working model of the Solar System where the planets move at correct relative speeds around the Sun.

10 POINTS

Meteorite

Occasionally some meteors (page 21) are large enough to land on Earth. They are very old and tell scientists about the time when the Sun and planets were formed.

10 POINTS

Index

Answers: P9 Polaris – Pole Star; **P11** Constellations – 88; **P12** Orion – The Hunter; **P15** Jupiter – Galileo; **P22** Great and Little Bears – the Bears have tails; **P25** Mizar – it is a double star; **P36** Sirius – Dog Star; **P50** Hercules – 12 Labo

i-SPY How to get your i-SPY certificate and badge

Let us know when you've become a super-spotter with 1000 points and we'll send you a special certificate and badge!

Here's what to do:

✓ Ask a grown-up to check your score.

✓ Apply for your certificate at www.collins.co.uk/i-SPY (if you are under the age of 13 we'll need a parent or guardian to do this).

✓ We'll email your certificate and post you a brilliant badge!